EVOLUTION vs. CREATION

THE FINAL WORD

by

Z. RICHARD SAWAN, M.D.

authorHOUSE

1663 LIBERTY DRIVE, SUITE 200
BLOOMINGTON, INDIANA 47403
(800) 839-8640
www.authorhouse.com

First published by AuthorHouse 06/08/04

ISBN: 1-4184-6148-2 (e)
ISBN: 1-4184-4970-9 (sc)

Printed in the United States of America
Bloomington, Indiana

This book is printed on acid-free paper.

DEDICATION

To my lovely wife, Mimi, without whom I wouldn't have
been able to figure out my personal computer

To the honor and glory of God's Holy name

IN GOD WE HAVE TO TRUST

PREFACE

Like everyone else, I watched in utter horror and disbelief the events that unfolded on September 11, 2001. It was indeed a terrible testament to the evil and callous nature of modern man. It was also a terrible day not only for our nation as a whole, but also for every decent, God-fearing person all over this world. As Christians, we preach love for not only our neighbors, but also for our enemies, or so Jesus Christ taught us. It definitely was a physical and spiritual 911 wake-up call, as far as I was concerned. How could God have allowed something like this to happen to us? Where was God when this event was unfolding? As one patient of mine stated very angrily, "Why didn't God take hold of those planes' controls and bring them down safely?" That patient left my office a very disappointed man.

Several days after seeing those gut-wrenching images over and over again, I slipped back into consciousness and a realization of how fragile our lives really are, and how uncaring humanity is as a whole. I have also come to know just how far man will go to accomplish his own desires and will "in the name of God." People use God to accomplish their desires, rather than let God use them for His. Like most of you, I sought refuge and solace in my God, His Son, Jesus Christ, and His word, the Holy Bible. September 11, 2001 was such a worldwide, cataclysmic event that I believed that such an event should be predicted in the Holy Bible. If it were in the Bible, I was going to find it. It had to be there. I even recalled reading somewhere in the Bible about a falling tower. Thus began my reading and search through the Holy Scriptures to try and understand and prepare myself, not only for the world to come, but also to try and find answers as to why the world was in such a chaotic state.

What I found in the Scriptures would change me as a Christian and a scientist forever. I found answers to questions I had had for many years, puzzling and difficult questions that were not only conflicting as far as I was concerned, but also challenged the very essence of my faith. For years, I have struggled with my own Christian beliefs and the overwhelming evidence that scientists and evolutionists had proposed for the past century and a half. It looked like I had finally found some real answers.

Not wanting to seem boastful, but I've "read" the Bible several times; still, as I started through Genesis again, I realized that I was reading and interpreting the Bible in a manner I never had before. I began to see and interpret things in a totally new light and with an unusually different perspective. How could I have missed the events in Genesis all this time? Prior to this, I would read the Holy Bible as if in a trance, accepting and feeding off my previously stored thoughts and beliefs. My mind interpreted the meaning of the Bible verses before I even read them. I was brainwashed because of my prior learning in Christian doctrine. I was taught not to question, but to be a good Christian. Besides, I was in no way capable of understanding the Scriptures. If I had any questions, I had to go to my priest, and he would decipher them for me. That is what I call total mind control.

I never found the destruction of the World Trade Towers predicted in the Bible, but I did find something else. In fact, this event pales in comparison to world events later to come, predicted in the book of Revelation, as well as in several other books in the Bible, which are soon to affect us all. Many people are afraid to read the prophecies pertaining to our era, but I assure you that the old adage "what you do not know will not hurt you" does not apply in this instance. What I did find (which I believe is not of my own wisdom and revelation) in the beginning chapters of Genesis compelled me to write this book you are reading right now. It was such a different way of looking at and interpreting the Biblical verses that even I was surprised.

Now you have to understand something about me. Other than schoolbooks and medical literature, I have never read a book in its entirety in my whole life. When I was younger, I remember that I was forced to read a book a week, which I totally resented and tried to avoid. I was able to get away with it by cleverly fooling my teacher. I never did read one book a week, but rather read enough to convince my teacher that I was doing what he wanted on a regular basis and with the utmost enthusiasm. Up to this day, I still don't read novels and books like other people do. My excuse now is that I am too busy with my medical practice and family, and that I just don't have the time, and that's true! With work and family obligations, there is no time! However, I am sure that if I were to search my subconscious mind, I would find out that this really goes back to my old professor many years ago when I was only 11 years old. I guess it is psychological. Now, I have written a book. I can't believe that this is happening. Until September 11, 2001, if someone had told me that I would write a book, any book, I would have laughed at that possibility.

The interpretation of the book of Genesis, which was like a breath of fresh air, is so different and compelling that I felt that I had to share it with everyone. At first I could not believe what I was reading and how I was interpreting it. With a background in science, I was starting to see the hint of evolution in the first two books of Genesis; in fact, I saw it even in the first few sentences. I then began to research evolution and in doing so, I was able to connect the two. This secret was there all the time and in the very first chapters. The explanation of the evolution of animals and plants proposed by evolutionists was now very easily explained. Although the Bible did not go into a lot of the details (in fact, I think that it was really skimpy), that hint of evolution was there all the time, and it was a very real revelation. The discoveries of man pertaining to evolution fit into the biblical version like a hand in a glove. There was no mistaking it.

However, when it came to man, I was almost convinced that man had evolved just like the other life forms on this planet. After all, scientists had a pretty convincing theory. Both the Bible and the scientists were in agreement with respect to the evolution of plants and animals. Which theory was correct—the Bible's creation story, or the scientific theory of evolution? I had reached a dead end! I was writing this book to convince readers that man was created, but I was stuck. For a few weeks, I was walking around in a trance, like a zombie. Everything that I was taught, everything that I had ever learned was open to question. I began to question humanity's existence on this earth. I began to question my very own existence and purpose in this short life. In my soul, I believed that man was created in the image of God, just as the Bible says. Had I been wrong all this time? Did I overlook something?

After I had awakened from that trance, I went right back to the pages of the Holy Bible. The answers to my chaos had to be in there. They had to be! I then came across many hidden secrets in the book of Genesis, hidden in plain view. It was right under my nose. I could not believe it! The Bible had not changed; it was the same. What was different was that I had changed. Rather than reading the Bible influenced by my previous memories and interpretation, I was reading it with a clear mind, open to any interpretation, and a totally objective viewpoint, the way I was taught as a physician. I had to clear my thoughts of all and every bias. This will be revealed to you as you proceed to read this book. It has made a change in me, and I believe it will also make a change in you. It not only strengthened my belief and faith in God, but also reinforced in me the awesome power of God, who is the source of all power and life in the whole universe. We all have learned from our youth, through different religions, "our" interpretation of God. Unfortunately, "our" interpretation of Him is someone else's interpretation—a religious leader's or teacher's. For the most part, we have become misguided with a false and incomplete vision about whom God really is. Many of us are worshiping a God whom we don't know and understand. That misunderstanding has made man the most dangerous living being on this planet. And that is where humanity, as a race, stands today. However, I think we all agree that He is all-powerful; that is, if we still believe in Him.

I don't think that those terrorists knew or believed this fact because they wouldn't have claimed that they did what they did for God. He doesn't need any of us to do anything for Him! He doesn't need our "help" with anything! Believe me, He can handle things just perfectly, even though you don't agree, or can't see the purpose of His work. If we, as a world community, know who God is, then we wouldn't do the terrible things we do to each other. We also would not be in the mess and chaos we are in today. We have become worse than animals, and yet we consider ourselves the most advanced evolutionary species. I do not think so! If we were created in God's image and likeness, something happened long ago that made us worse than the animals and took us further and further from the image God wanted to express in us, the wonderful expression of God the Almighty. We presently live in a world without the true God, and as you can see, the end of this path will be total destruction and damnation.

As you read this book, please do it with an open mind. I know that many would stop here because <u>I have expressed my belief in God and creation.</u> <u>However, I also believe that evolution occurred</u>. To understand this "contradiction," you will need to read on. And as you read on, you will find out that there is no contradiction in my belief, which is derived from the Bible, thereby erasing any contradictory misconception in your mind that may exist with respect to Genesis and creation. I hope you enjoy this book, and I also hope it inspires you to search for the real God, who is still alive, by the way, and to see all around you His awesome power and glory. If you believe and trust in Him to resurrect you after death, shouldn't you believe and trust in Him while you are alive? If you believe that He is so powerful, don't you think that He would have something to say with respect to the Holy Bible, a book that represents Him? He is not dead, and He has a lot to say. He talks to us every day in different ways; however, if we are not paying attention, we are looking the other way. If we are not listening to some "blind man's" interpretation of the Holy Bible, we are listening to our misguided thoughts and beliefs. Listen up! In God we SHOULD trust!

CHAPTER 1
<u>EVOLUTION OR CREATION</u>

What opposites! What a fight! What a rift in ideologies and beliefs! What is the truth? Did we evolve, by chance, out of nothing, as many diverse peoples and nations and scientists think? Or were we created for a special purpose by a Supreme Being—God—like others believe? Or did we originate and transport ourselves in some form or the other by UFOs from some distant galaxy, and then plant ourselves here as some superior alien race? Who and what are we, and where did we come from? Where can we attain and decipher knowledge about our origin and past? Is it even really that important? Does God even exist, or did He exist previously, but is now dead? When can we finally answer these questions and put them to rest? There is definitely a great divide between evolutionists and creationists; one, which to many, appears uncross-able. The battle is a ferocious one that sometimes ends only with blows and name-calling, leaving nothing accomplished. It's difficult for evolutionists to convert creationists and bring them to their side, and vice-versa. Some evolutionists and scientists make light of creation, calling it a "story," thus making them atheistic, lukewarm, or even cold in their religious faith or background Biblical scripture is a way, as some claim, for world religions and leaders to try to control and manipulate humanity with fear.

The behavior and attitudes of present-day religions and their so-called leaders do not make it easier to believe that there is a Supreme Being. In fact, we have lost faith in both God and religion as an institution. Some of these religious leaders behave and think worse than animals, thus causing one to think that maybe we really are "descendants of the apes." By using this comparison, I am probably insulting the apes. Their behavior inspires one to turn away from religion completely. Just look at the recent crisis within the ranks of the Catholic Church, and how they approached the issue of sexual abuse against their loyal and helpless followers, mere children at the time these awful crimes were committed. They, meaning most religious leaders, and representing all religions, are an embarrassment for what they are supposed to represent. In fact, their ways and thinking reflect their own ideologies, and not that of their religion or their God. They all seem to have the same God, but yet they are divided in their beliefs, even within the ranks of their religion. Let's face it; these people, or so-called leaders, seek their own interests and theirs only. No wonder the masses are deserting God and traditional religion and seeking alternatives that spiral them into what I call total emptiness.

For those on the other side, they can't even begin to explain the discoveries of the scientists, accepting that what they believe might be right or wrong as long as it is in coordination with their own beliefs, or what they have been taught or brainwashed to believe since childhood. They simply walk around in blind faith, totally ignoring the evidence that the earth has to reveal, believing in so-called religious experts or blind religious organizations. I used to be one of them. Some are so ignorant that they refuse to even examine or read the evidence. The issue of our origin is so argumentative and crucial that it needs to be examined, and it is one that needs an explanation. It is a matter of life and death! What I am about to reveal, hidden in the Holy Bible throughout the ages, and corroborated by scientists and evolutionists, will open one's eyes and make those who are cold or lukewarm in Christ explode with exhilaration and fervor to admit the existence of God and pursue their own salvation through His Son, Jesus Christ. For those atheists, it might open the door to their creator and their own salvation. It might at least wake you up, and encourage you to "seek and you shall find."

First, I am going to give you some history and background on the man who started it all, Charles Darwin (I shall show you what he did right and where he went wrong later on in this book). I am then going to discuss a very interesting theory, maybe even a fact, called The Big

Bang, which is a theory that possibly explains the origin of the universe, or at least part of it. I shall then examine both opposing views of evolution and creation and what our beliefs are today. Next, I shall then link both opposing factions through the amazing secret truths hidden in the Holy Bible, and the discoveries made by archeologists in the past decades. These revelations you are about to discover will indeed set you free from the bondage that has deceived many for centuries, and give you a freedom you can't imagine. You have no idea what is in there. These truths have been hidden in plain sight for centuries. Is prehistoric man the image of our past, or were we created in the image of God? I shall once and for all answer the question which we have all asked ourselves over and over for the past century and a half, since Darwin introduced his theory of evolution and confused us all: <u>WERE WE CREATED, OR DID WE EVOLVE?</u>

CHAPTER 2
<u>EVOLUTION</u>

To the scientist, the earth is an open book. Its past history can be found buried within its layered strata and gravesites. The bones and imprinted casts of preexisting creatures found in the heart of the earth read like a storybook; however, there are a lot of pages missing in this puzzle of the ages. For decades, scientists have examined the fossils buried within the earth for millions of years and have concluded that the process of evolution did occur and <u>THAT EVOLUTION IS A FACT</u>. I welcome you to read that "storybook" as well. Man has also examined the universe: its light, gases, meteors, etc., to further search for explanations on the origin of life. Man is still studying the universe, spending billions of dollars sending probes into the vastness of space to try to analyze planets, meteors and comets. We also keep our "ears" open, hoping to find the existence of life outside our solar system in order to finally prove that we are not alone. We have spent billions of dollars in this quest, and we are prepared to spend billions more to find the truth. That would be powerful ammunition for the scientists and atheists to disprove the creation "story," and thus throw God out of the equation. Some are actively seeking to prove that God doesn't exist.

According to scientists, the earth and the universe have been in existence for billions of years. They have found the presence of life that existed millions of years ago in the numerous unearthed fossils recently discovered in the past century. The age of these specimens is determined by radioactive carbon activity (one method of determining age of fossils) left in these specimens. As with everything, there is some controversy with respect to the numerous methods of dating specimens and their accuracy. After examining numerous specimens, man was able to come up with an autobiography of the earth, as it were. Please refer to *figure 2*, which outlines the man-made timetable of the earth's past, and the appearance of plants and animals through its prehistoric existence based on this radioactive carbon dating. I am not here to validate or reject this method! I am not an expert in this method of dating specimens, and I am not qualified to direct you as to which method is best. With respect to the different methods of dating fossils (their accuracy as well as the controversy surrounding them), I am very happy and willing to leave that to the experts. Let them fight it out on this matter. As you will see later, this will become totally irrelevant as the truth becomes revealed. I will show you the sequences of the appearance of life, but again, I shall leave it to the experts to fight and argue on the timeline of the appearance of that life (*figure 2*).

But how did life begin, and what part did God play in it, if any at all? Did everything evolve, or was everything created? Was man created special, while the rest of life on earth, including plants and animals, evolved? According to evolutionists and scientists who were able to create DNA (deoxyribonucleic acid, the "genetic blueprint," see *figure 3*) in the lab by simulating conditions of the earth billions of years ago, that's exactly what happened (the basic blueprint of life was formed by chance without any divine intervention). A primitive form of genetic code was formed in the violent earth milieu, which later, after millions of years and with a lot of luck, became a living life form.

From that point forward, the evolutionary process began, became more sophisticated, and, over hundreds of millions of years—voila! We are now in the twenty-first century with the modern-day results of evolution. The basis of all of these amazing changes is the result of beneficial mutations that have occurred over millions of years. Given enough time, according to scientists and evolutionists, anything can happen. Now, however, even evolutionists admit that the chance of a beneficial mutation is 1:10,000,000, to be extremely generous. Now think also—what might be the odds of that first life-form (which has undergone a beneficial mutation) surviving in such a hostile earthly environment? If it did survive, the odds of another beneficial mutation are 1:10,000,000. What, then, are the odds of millions of more beneficial mutations in that species to reach the complexity of a living being like man? Add a lot more zeros after the one. What, then, are the odds of thousands of species forming, all undergoing their own individual beneficial mutations, to become such complex creatures that we have today, most of which not only live in symbiotic relationships with other animals, but also with plants and nature? I remember learning in college about a principle in either physics or chemistry that simply stated that everything in the universe goes from a state of order to one of disorder. Why is it then that evolution on earth didn't obey that principle?

One can see the complexities amongst these relationships by just watching any channel on plants and animals, as well as relationships amongst animals themselves. I was watching the Discovery Channel one evening about the turkey and king vultures. The turkey vulture evolved with a keen sense of smell, which the king vulture lacks. It can smell a rotting carcass from very high in the air, even if a canopy of trees covers the carcass. As the turkey vultures approach the carcass, the king vulture, which knows their intentions, follow directly behind. The turkey vultures that get there first have to wait for the arrival of the king vulture so that it may tear the tough hide of the carcass with its sharp beak, an attribute which the turkey vultures lack. This way, both can feed, and thus survive. Isn't that amazing? This is but one example of the complex relationships that exist between all living things. There are millions of other complex relationships. Wow!

So what are the odds? Add a lot more zeros; I mean a lot more. What's also impressive is that all this is occurring on this tiny earth with just the right atmosphere, and which happens to be the perfect distance from the sun. I don't have enough space in this book to print the infinitesimal amount of zeros after the number one. What you have here is a theory that is just a <u>BIG FAT ZERO</u>, built on a lot of hope, chance, wishful probability, and individual speculation. <u>OR IS IT</u>?

THE MILLENIA OF TIME

Numbers are in millions of years

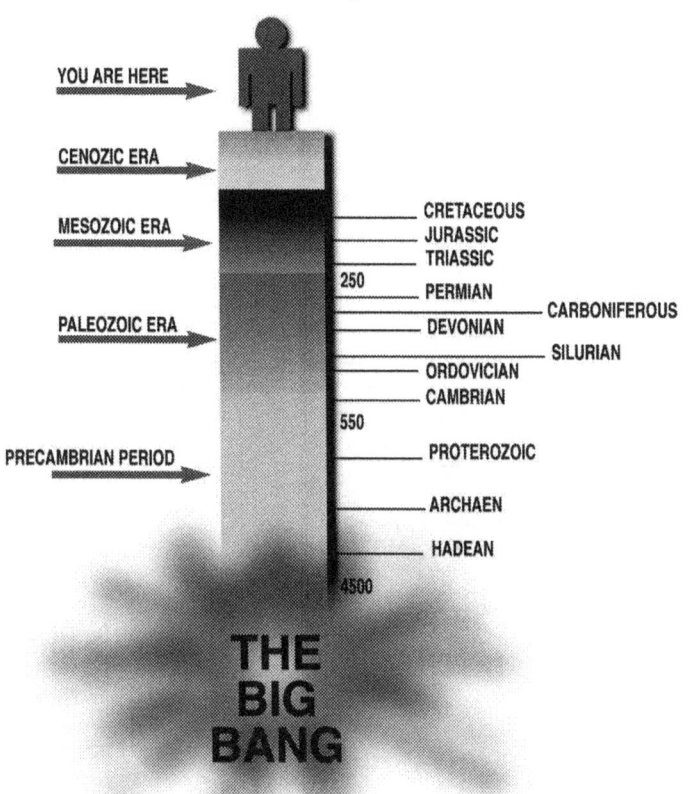

FIGURE 2-THE MILLENIA OF TIME.
Numbers are in millions of years

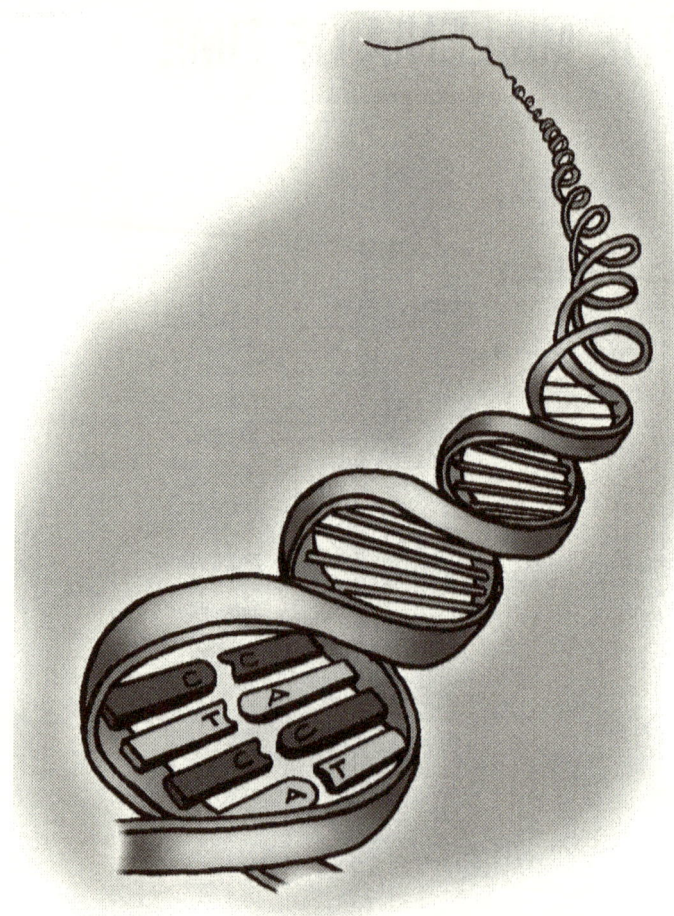

A-Adenosine T-Thymine G-Guanine C-Cytosine

THE HELICAL DNA MOLECULE
THE BLUEPRINT OF ALL LIFE
<u>FIGURE 3</u>

CHAPTER 3
<u>CHARLES DARWIN, THE</u>
<u>FATHER OF EVOLUTION</u>

FIGURE 4-Picture of Charles Darwin

Charles Darwin was born in England in the year 1809. He was born into a family of wealth and education and enjoyed all the privileges that came with it. Of notable mention was his grandfather, Erasamus, who was an expert in botany, the study of plants. He, Erasamus, was a big influence on the life of young Charles, especially on some "wild" ideas on evolution. These ideas on evolution were published at the time, but were dismissed as too radical or outrageous. Nevertheless, the young Charles Darwin was heavily influenced by his grandfather and had access to these books. He therefore had the opportunity to read them all, and thus the seed of alternative thinking was planted.

Being from a well-to-do family, Charles was exposed to proper schooling. He even went to medical school, but because of his dislike for it, he decided to quit after two years (he probably wouldn't even have considered this if he were alive during these tumultuous times in medicine we have today). After Charles quit medical school, his father influenced him to go to Cambridge to become a clergyman. By the way, he was not a spoiled brat. It was during that period of his life that he pursued numerous different theories in science by reading many books and manuscripts written by the influential authors of that era. He probably wasn't satisfied with the biblical explanation of the origin of life, which explained the pursuits, passions and dedications in his life. Thus, the seed of evolution, which was planted by his grandfather, was nurtured and watered by the present-day thinking of that time.

After graduating from Cambridge in 1831, Charles Darwin embarked on a long oceanic voyage to find and document the existence of new species. This journey would cost Charles five years of his life and would lay the framework to help nurture the little seedling, which, after a few years of growth, would rock the very foundation of the scientific and theological world. During his travels, which eventually took him to the Galapagos Islands, he not only recorded the existence of a plethora of new species of animals and plants, but he also documented numerous fossils of both. It was, indeed, a thrilling and fantastic voyage of discovery.

It was not until 1837-1838 that he began to look at his collection in a different light and thereby started to entertain the theory of evolution based on the evidence he had, notably the theory of natural selection. Simply put, this theory suggests that the environment, which is under continuous change and never stagnant, can select and interfere with existing species, plants and animals, which obviously depend on the environment for their very existence. So, as life unfolds, favorable characteristics are selected out, and unfavorable ones are "voted" out. Subsequent to this, one would either see a change in an already existing species, or the death of a species, or the birth of a new species altogether, depending on the mutant gene. So, just like the environment we live in, all species of life are in a state of constant change. Keep in mind that these changes occur over hundreds of thousands to millions of years. Darwin was on to something really big.

This theory implies that the differences in species in nature are a result of natural genetic selection and variation, and are not caused as a result of species trying to change themselves to adapt to the particular environment in which they happen to live. Plants and animals cannot just will themselves to change. This theory strongly implies that these differences are definitely not a result of a divine creator, God, but are instead random events, naturally occurring in a very labile environment in which the organisms are presently living and competing to survive..

While writing his book, *The Origin of The Species*, Darwin received a manuscript from a prominent naturalist by the name of Wallace. In his manuscript, Wallace asked Darwin to comment and opine on his (Wallace's) idea of natural selection, which was the very meat of Darwin's book, half of which was already written. What a shocker! This was indeed a big surprise and probable disappointment for Darwin as all those years and hard work went down the drain. He overcame this by meeting and agreeing with Wallace to simultaneously publish their books. Darwin, however, was to make the bigger impact of the two. His book was published near the end of 1859. The appearance of his book, which contained such radical views, caused a major uproar amongst scientists and theologians (strong creationists at that time), who, before 1859, existed in "perfect" harmony. Just imagine the debates, the name-calling, and controversy on the heresy of the ages. I would think that there were probably threats on Darwin's life. He may have had a few bodyguards. There was both praise and ridicule for Charles Darwin.

In 1871, Darwin published his book, *The Descent of Man*, in which he focused on the origin of man. In that book, he implied that humanity descended from a "pre-ape-like" creature. What a time it was. If the first book created uproar, just imagine what his second book did by suggesting and implying that man descended from the apes. Nothing has really changed since then as we enter the twenty-first century—that is, until you read this book. There are still heated debates and inconclusive arguments going on all over the world. People are on one side or the other. The knowledge we think we know, and man's stubbornness, keep it that way today and don't allow any compromise. Well, there is a better way, and I hope this book will show you and help you understand the truth on the origin of all life in its entirety, as we know it today.

At this point, I must make a few remarks about some of Darwin's theories in the aforementioned books. I have a lot of respect for Charles Darwin for doing what he did in the period in which he lived. In fact, some of his theories significantly help support this book and my theories. I know that some of his theories are true and without them, even the "story" of creation would be in question. Some of his theories actually make the "stories" of Genesis real. However, even though I can see how he made some of the conclusions he did on man, I believe he missed the mark with respect to the origin of man, as many before and after him have done. He was only partially correct. He applied the same evolutionary principles of non-human organisms to man himself. As you continue to read this book, you will understand what I am trying to say.

Man is truly a distinct and separate species from everything else that has ever existed here on earth. Whereas Darwin used the basic structure common to man and apes, for example, to show that there is some ancestral connection, I say that those similarities are part of a basic framework to help the organisms function on this planet. It's not unlike two separate car manufacturers producing two totally different cars with respect to shape, speed and luxury; yet, the basic essence in design and structure are very similar. So even though they may look and function similarly, they both had different "creators." One can say the same about houses, telephones, airplanes, etc.

This distinction of direct creation by God applies primarily to man, although some of the Garden of Eden animals were created directly by God. The main focus with respect to the last statement, though, is the creation of man. However, as you will learn later, this basic evolutionary thinking which Darwin proposed does apply to plants and animals, including prehistoric man; still, I disagree that it applies to present man. In actuality, Darwin was partially right, and he was really on to something. But again, it did not apply to man. Notice that I did not include man as an animal. Even though man may look like an animal on the outside, his inner soul is far from it. In essence, God put a living soul in a physical body. Confused? I bet you are!

It was at this point in writing this book that I also became confused myself. I alluded to this in the preface. I was totally devastated because, at one point, I couldn't refute some of the theories Charles Darwin put forth in *The Descent of Man*. And not only that, but the Bible supported some of his book's conclusions with respect to the evolution of plants and animals. As I said earlier, the whole purpose of this book is to convince readers that creation did, in fact, occur. I shall prove that evolution occurred with respect to plants and animals, but where was the proof that God created man? There had to be some proof, direct or indirect. At that point, even I was questioning the very existence of God, as well as the purpose of the human race. I then saw the light I was looking for in the book of Genesis, and I shall point it out to you soon. So don't get discouraged, because all of this will be explained later in great detail and brought into the light. You will not be disappointed.

CHAPTER 4
<u>CHARLES DARWIN'S THEORY</u>
<u>ON SURVIVAL OF THE FITTEST</u>

Can this theory be so far-fetched and outrageous? I used to think so. Many years ago, I took a course in evolution, hoping to prove to myself once and for all that Darwin was misguided, deceived and totally self-centered. At the end of the course, I totally agreed with Darwin on his theory of the survival of the fittest. I was the one who was misguided and deceived, a hostage as it were, by my own religious thinking. However, I couldn't come to grips with the belief that life was an accident or a chance event. As a physician, I have seen the existence of this theory that, in my mind, casts no doubt on the plausibility of Darwin's thinking. To refresh your memory on what this theory proposes, please read the previous chapter.

Let's look at the first example. Malaria is endemic in several parts of Africa. In addition, several Africans have the disease sickle-cell anemia. It has been observed that some Africans get infected with malaria, while others have become resistant to the disease. To simplify, let's assume that persons without sickle cell anemia and who have normal hemoglobin in their red blood cells carry the genes <u>ss</u>. Those with sickle cell anemia and, therefore, abnormal hemoglobin in their red blood cells carry the genes <u>SS</u>. Individuals with <u>ss.</u> who marry those with the <u>SS.</u> genes will have offspring with the genetic combination <u>Ss</u>. These are the sickle cell trait generation that has become resistant to malaria. Both <u>SS</u> and <u>ss</u> individuals are susceptible to malaria. If man doesn't interfere by providing medicines to fight malaria, then a lot of individuals carrying the S̲S̲ and s̲s̲ genes would die off, leaving the S̲s̲ individuals. This is natural selection. The environment has naturally selected a genetic combination, which is beneficial to that particular species, mankind, allowing it to survive and thrive in that particular environment.

How about another example? In the mid-twentieth century, Europe's air and forests were heavily polluted. At that time in Europe, the industrial revolution was in full swing, and with it came a lot of uncontrolled pollution. Thus, in some areas, entire forests were covered with soot. In the English forests lived a species of moth, both dark ones and white ones. With soot on the trees, the white moths stood out very clearly and were gobbled up by their predators. Prior to the pollution of the forests, there were equal amounts of dark and white moths. Now, because of man's manipulation of the environment, an artificial form of natural selection had occurred. You now had more dark moths as compared to white moths. It was that simple.

A much more rapid transformation in natural selection occurs every day in bacteria. If one exposes bacteria to antibiotics, one can see the emergence of resistant strains of bacteria to that particular antibiotic. The antibiotic doesn't cause the bacteria to change, but it allows the selection of those bacteria that possess the ability to resist the antibiotic to remain alive, and thus perpetuate the genetic information to their offspring. We are now seeing bacteria that possess resistance to several antibiotics.

Let's look at another example. Assume you have a herd of animals grazing in their natural habitat somewhere. Now let us assume that a drought strikes the land, making it difficult for the grass to grow, and thus diminishing the food supply. The only food source left would be the vegetation on trees, which are able to tap into the ground water supply with their long roots. Now just suppose that in that animal species, there is a variation in the neck length and height. What do you think would happen? As the animals compete for food, only the ones that are able to reach the leaves on the trees would survive. This would naturally select out animals within that group that are taller, as well as those that have longer necks. That's natural selection in a nutshell.

Don't these examples make sense? Darwin saw creatures that were similar to one another, but because of isolation, time, and a changing environment, they developed into separate species over thousands of years. Organisms can change naturally, and this is a fact that you, the reader, have to keep in mind as you continue to read the rest of this book. Organisms don't change because they feel like changing.

CHAPTER 5
<u>THE BIG BANG THEORY</u>

The Big Bang theory is a simply put theory with many hidden and amazing complexities that try to explain the origin of the universe. These words just don't express to you, the reader, the amount of time and effort in research that was accumulated by scientists to mostly agree to this theory. The Big Bang is the culmination of over 150 years of hard work by the major scientists of that time period. The Big Bang theory is a result of numerous and countless formulas and observations which I myself do not understand. Needless to say, some scientists are still busy working on it. On that note, I don't think I can do them any justice in trying to summarize their efforts in this paper with just a few words, but I shall try. I am also not in any position to refute or accept this theory. What I am trying to do is to link this theory or any other with the events played out in Genesis in order to try and explain what possibly happened in the beginning, as described in the book of Genesis. Is there any similarity between what the scientists believe and what is written in the Bible, with respect to the origin of the universe? Could it be possible that something like the Big Bang occurred billions of years ago, and could it even be mentioned, if subtly, in the Holy Bible—so subtly as to go unnoticed?

Drawing of color spectrum
Red -orange- yellow- green- blue- indigo- violet.
<u>Figure 5</u>

When a heavenly body is "motionless," it emits a certain color of light. When it is moving away or toward the observer, that color of light changes, so it is easy to tell whether an object is moving away from you or coming toward you. Scientists have observed that when a celestial planet or object moves away from earth, the "observer," the color of the light waves in the visible spectrum of light shift towards the red end (*the Red-Shift*). Conversely, if an object were moving toward the earth, the color of light would shift to the blue end. This is fairly consistent and predictable. It was observed that the majority of heavenly bodies were moving away from the earth. Why? Did all of these heavenly bodies break up from a central unit at one time? And why were they being propelled away from each other? Why is that?

Because of this and other evidence, many scientists, in a nutshell, have therefore concluded that the universe is expanding at a very rapid rate, and has been doing so for about 15 billion years. By estimating the distances between the earth and some well-known objects in space, they are able to "guesstimate" the age of the universe, or the birth of the Big Bang. About 15 billion years ago, there was a gaseous cloud or star or nucleus that exploded with tremendous force, sending particulate matter, debris, gas, or the contents of atoms in all directions at tremendous speeds. We who are alive on earth today witness these planets and galaxies moving away from us as shown by the *Red-Shift*. It is one piece of evidence about our past.

Throughout the years, these "gaseous" particles first liquefied and then solidified to form the complex planetary systems we know today as the universe. The earth presently has a central, liquid, molten core covered by a thin, solid crust. This liquid core began to cool down at the periphery to form this outer crust, as we know it today. This solidification took millions to billions of years. We observe planets that are made solely of gases, while others appear earth-like, but are without life. One can see fiery planets today being formed in front of our very eyes, which probably gives testament to our past. One of those, the Sun, gives the earth energy to sustain life. Was this an accident or a planned event?

The Big Bang theory doesn't stand without criticism; in fact, there are many scientists and creationists who do not accept this theory at all. One example, for instance, that was used to refute this theory was the discovery of super clusters of galaxies millions of light years away, and separated by millions of light years from each other. These galaxies, in turn, were millions of light years long and millions of light years wide. It is estimated that the time necessary to create such an enormous galactic system would require approximately 80 billion years. Another later discovery of a system of galaxies would have required 150 billion years to develop. This system was also very similar to the one previously described. According to Big Bang proponents, the origin of this huge cosmic explosion occurred about 20 billion years ago. This, by no means, can begin to explain the existence of the super cluster of galaxies seen throughout the universe.

Could it be possible that both competing groups might be partially right? Is it possible that there were several Big Bangs? Could the origin of the universe have different origins altogether, including the Big Bang, which may have formed our galactic system, as well as others nearby? Why are scientists and the human race so polarized when it comes to theories and opinions that they are not willing to consider a middle ground, which, in fact, may solve their dilemmas and predicaments, as well as answer their basic questions? I must assume that it has something to do with ego.

Does the Holy Bible give any confirmation that the Big Bang, or something like it, actually happened billions of years ago? At this point, it is not important to me as to the exact number of billions of years. I shall let the scientists fight that battle while I watch and listen on the sidelines. The real question is, was it a possibility? If it did happen, I would like to reverse the tables on these intellectuals and ask the questions. Do the discoveries of the scientists confirm the validity of the Bible, or at least the "story" of creation in Genesis? And if they do, do they make the Bible more of a product inspired by God rather than men, though it was written by men? And if so, would those scientists be more apt to listen and take another look at the Holy Bible?

One would be very surprised indeed to know that the Bible describes such an event, but not in such detail to make it very obvious. The Bible is notorious for omitting details, which, by itself, can make a stumbling block for many. That is why it's been hidden from view for all these generations. Also, the interpretation of such an event in the

Bible would have been impossible prior to the suggestion of the scientists of such an event. Because we know what this theory states, it becomes easy to see it unfolding in Genesis. The more and more scientific information that we, as a human race, attain, the more we are able to unearth the hidden secrets concealed in the Bible under our very noses for all these centuries. Are you surprised? So was I. It took the discoveries of archeologists and the theory of evolution to unlock the hidden truths in Genesis, as well as the hidden truth on the origin of life here on earth, and the origin of the universe. Open your mind, and read on!

CHAPTER 6
<u>CONFUSION IN GENESIS</u>

According to modern day scholars and experts, at least self-proclaimed scholars and experts, the first two chapters of the book of Genesis provide us with two different story versions of creation. There is apparent difference in the order of the appearance of created organisms, including man, in the two stories. This has led scholars and biblical experts to conclude that both stories are derived from different sources. In fact, these two sources are called the Priestly and the Yahwistic traditions. The former is represented by Genesis 1.1 to 2.4, while 2.4 to approximately 2.25 represent the latter. It is believed that the author(s) of these two traditions represented different social structures whose ideas are reflected in their themes.

"Genesis is thus a syncretic account whose inconsistencies are especially visible when the events of human creation are described," writes Ewa Wasilewska in the magazine *The World and I.* So not only is the scientific source of the origin of life confusing, but the ultimate source, The Holy Bible, appears to be inconsistent. Or does it? This is a pretty strong statement by Wasilewska, which probably represents the same thinking of the scholars and scientific community. But does it hold water? In addition, it is believed that since the older Yahwistic tradition follows the younger Priestly tradition, the authors of the Yahwistic tradition had to make adjustments in their account to show some similarity to the account of the Priestly tradition.

The Priestly tradition, as the name implies, represents the ideas of authors totally embedded in the religion of the

Israelites at the time it was written. One first notices the – seven-day story of creation that patterns the seven-day week of the Israelites, the seventh day being a day of rest. Since God rested on the seventh day, man is ordered to do the same on the seventh day of the week, the Holy Sabbath. What better way to control the masses, as many would conclude. As one continues on in the Priestly account, the second thing one notices is the orderly account, step by step, of the creation process. Plant and animal life were created before Adam and Eve, and, according to the Bible, the latter was to rule and control the former.

On the other hand, The Yahwistic tradition, as the name implies, focuses primarily on the God of the Israelites, Yahweh. In this account, man was created before the animals, which differs from the Priestly account. Eve was later formed from a rib of Adam. The central focus, according to biblical scholars of the Yahwistic tradition, is the involvement of Yahweh in the creation of the male and female humans, and the superiority of the male over the female. Of note here is the fact that both animals and man were formed out of clay, but woman was not. She was created out of a rib of Adam. To many, this reads like a storybook.

It is therefore not surprising that a lot of confusion emanates from Genesis with respect to the origin of life. It is totally understandable. If one doesn't know the Bible, then it would be easy to conclude that the discrepancies in Genesis were created in Genesis itself, and not by modern-day scholars. Does the word of God, from the very beginning of Genesis, start off with contradictions, confusion, and inconsistencies? If that is true, how can we rely on what is to come later in the Bible, whether it is in the Old or New Testaments? <u>This is an important and critical question</u>. If these apparent contradictions in Genesis cannot be thoroughly explained, then our Bible, the whole book, is worth nothing.

What you are about to learn in this book will change the way you look at the Holy Scriptures. THERE ARE NO DISCREPANCIES OR CONTRADICTIONS IN THE CREATION ACCOUNTS OF GENESIS. And I shall soon prove it to you! By the time you are through with this book, you will learn and see that truth as well.

CHAPTER 7
<u>CREATION</u>

<u>Figure 6</u>-Picture of Adam and Eve in the Garden
of Eden

According to present-day beliefs, it took God "six days" to create life here on earth, as we know it today. On the seventh day, God rested. Some say that a day was a 24-hour period, while others say that a day was 1,000 years. Some even think it was more. In essence, man and present-day life have been in existence for approximately 6,000 to 12,000 years, give or take a few, according to creationists. The scientists, or "the unbelievers," have no idea what they are doing, and their methods are obviously erroneous. Scientists may be correct in a lot of things, but when it comes to this religious issue, they are simply misled by Satan. They have really gone off course and missed the boat completely. On the other hand, the faith of the believers has completely blinded them from even looking at any of the evidence that the earth has to reveal...not just evidence, but facts. Why are we so closed-minded? Some of these believers don't even know what evolution is, and don't take the time to read any literature on the topic. To even entertain the theory of evolution would be labeled as dancing with the devil. It's amazing that in this age of information and technology, a lot of things are still taboo. There are numerous powers on this earth that try and keep us ignorant of many things, so therefore let's not give them any more help. Think free, and be free.

In Genesis, God <u>prepared</u> the earth to support the life He was going to create (after the second day) on the first two days of creation. Evolution proves that. He made sure that there was light and water, the two essential necessities of life, to support the living organisms He was going to create. He specifically listed, in order, the creatures that were made. I think that that was being pretty bold of Him to do so, unless it actually happened and can be later proven by man. Wouldn't you agree? But did He list everything? No, He did not. First, He made the plants, all kinds, on the third day. He then went on to create the sea creatures and the birds, which were <u>PLEASING TO HIM</u>. That was done on the fifth day. Then on the sixth day, or at least the first part of it, He made all kinds of animal life, some of which are still in existence today. Then came the creation of man toward the end of the sixth day. And God rested on the seventh day.

That, in essence, is the narrative history of creation that we were taught to believe. But is it the complete story? Was it that simple and straightforward? There is much more to it than meets the eye. According to theologians, there are two "stories" of creation. The first is the Priestly Tradition, which is described in the first chapter of Genesis. The second is the Yahwistic Tradition, described in the second chapter of Genesis. Because man doesn't know or understand the Bible, these two "stories" appear to conflict and set the Bible up to be labeled as contradictory. Even the religious leaders of our day do not have a good explanation for this possible contradiction. As a result, man has looked elsewhere to explain the origin of life, and even more so, the origin of humanity. The Bible was put aside, locked away and labeled a storybook.

But getting back to creation, God was probably really tired after creating the whole universe and all the living things on earth; or was He? Why did He have to rest? Isaiah 40: 28 says that He never grows tired or weary. Why did it take Him six days to create what he did? After all, He is God the Almighty and the Supreme, so, if He wanted, He could have made the world by just saying one word. This could have all been done in less than one second. Why did He give us a step-by-step account of the creation process? He could have created all of the plants and animals all at once. Wouldn't that describe a much more powerful and omnipotent God? If the Bible is the word of God, then what God said He did actually happened. If you don't believe that God exists, then the whole Bible is irrelevant, and just a bunch of stories made up by man to suit his own purposes. Does the earth have any record of creation? Has man found anything that comes even close to the creation account?

Were the fish and birds really made first? What have the scientists found in the archeological history of the earth? Do the discoveries of the scientists give any credibility to the Bible or vice-versa? Does the theory of evolution and the archeological discoveries show any similarity to what God recorded in creation in the book of Genesis? Where is the proof that God's word is true?

Who are we to believe? Scientists and evolutionists have found that the earth and its past living inhabitants have been around for millions to billions of years. On the other hand, creationists believe that all life has only been in existence for roughly 6,000 to 12,000 years. They believe this because that is what they think the interpretation of God's word, the Bible, says. Well, that is what their religious leaders have taught them. Does the Bible, in fact, say that? This faith has completely blinded them to the evidence available, making every other explanation of "creation" nonsense. The Bible tells us to prove all things. Don't you think that God could foresee the coming of this era, where a lot of information would be available to us? We are at a point in history when we can actually prove more things, including questioning and proving God's account of creation. It's judgment "day" for God! In addition, if one can prove the creation account correct, wouldn't it also prove that the Bible, though written by men, was inspired by God?

If you read 2 Timothy 3:16, you will see that "All Scripture is inspired by God." At the time these words were spoken, not written, the only Scriptures available were those of the Old Testament. John 10:35 says that what the Scripture says is true forever. Acts 7: 53, Galatians 3: 19, and Hebrews 2: 2 tell us that the laws and messages given in the Old Testament were handed down by angels, with a man acting as a go-between in some cases, and were shown to be true. Do we believe the New Testament only, and not the Old? "I don't believe the Old Testament because it's all a bunch of stories." Is this what you believe? If you believe in the New Testament, and that that Testament is a testament to the Old, then you either believe both or none. You cannot believe one without the other. The Bible is God's "spoken" words. Think about it.

Where can we go to find the right answers to all of our questions? Why, the Bible, of course, as well as in the recorded manuscripts of archeology. What I'm about to reveal is going to blow you away. The truth is indeed going to set you free, free from the chains that bind your mind. Remember, the purpose of this book is not to reveal the purpose of creation, but to reveal the truth about what actually happened. <u>ARE YOU READY</u>? Let's begin! Get your Bible (*King James*, *Good News Bible-Today's English Version*), and let's go step by step.

CHAPTER 8
<u>THE BEGINNING</u>

<u>Gen 1: 1, 2:</u> In the beginning
God created the heaven and
the earth. And the earth was
without form, and void.

The earth was <u>without form</u>. Many scientists believe that the formation of our universe began billions of years ago with the explosion of a large gaseous cloud, the Big Bang, which later became the present day wonder at which we now marvel (see Chapter 5). All planetary objects, including the earth, were in gaseous form, thus, <u>without form and without shape</u>. After billions and billions of years, this formless and shapeless gas collection became a planet with form or shape, thus giving birth to our present day earth. And the earth was <u>void</u>, meaning empty of life. There are billions of years between the words without form, and void. This coincides nicely with the Hadean and Achaean Eras. These eras were 4500 to 3800 million and 3800 to 2500 million years ago, respectively. You therefore cannot assume that if one verse goes into another that the time-period is "right away." <u>The Bible does not give everything in detail, but summarizes many events</u>. During those billions of years, parts of this gaseous form of the earth liquefied (to form water and molten lava) and later solidified to form our present day world. Some of it is still in liquid form, under the earth's solid crust, as molten lava. The evidence of our past is literally right under our feet. Few people would dispute that the earth is millions to billions of years old, but there are some that still disagree, making their only source of reference the Holy Bible. This is what I call sheer ignorance.

> <u>Gen1: 2:</u>....and darkness was
> upon the face of the deep.
> And the Spirit of God moved
> over the face of the waters.

It's obvious that during the solidification phase of the earth, a lot of water was formed. It was probably formed by violent atmospheric conditions, which, according to the scientists, existed at the time. The environment was indeed very inhospitable. Some Bibles use the term "raging" waters, indicating violent storms and, possibly, underground earthquakes stirring up the waters. So even after the solidification phase, the earth continued to be inhospitable. Without water, one cannot have life, and I believe that the Spirit of God was moving over the face of the waters (<u>Genesis1: 2</u>), guiding or observing creation in progress. The evolutionary process was being created and designed by God for reasons of which I am not entirely sure.

> <u>Gen 1: 3:</u> And God said, Let
> there be light; And there was
> light.

<u>The earth was in total darkness</u>. Light was allowed to penetrate through to the earth. Through what? If you leap forward to <u>Gen 1: 14</u>, day four, the lights in the firmament of the heavens, that is, the sun and moon, <u>were allowed to appear</u> or become visible to the naked eye. They were obviously made in the "beginning," as was the rest of the

universe, but the light of the sun was not able to penetrate through to the earth until <u>verse 3</u>. It is pretty obvious that the earth was in total darkness because it was blanketed by a thick atmospheric cloud or dust, which was formed as the earth was being formed, probably by severe volcanic activity, meteors impacting the earth, etc. I will let the scientists fight over that. Now getting back on course, it probably took thousands to millions of years for the light of the sun to appear in <u>verse 3,</u> and thousands to millions more for the sun and moon to become visible as that thick cloud cover began very slowly to settle down to earth. What this reveals is that a one-day period, as stated in the Bible, was not 24 hours or 1,000 years, but several thousands to millions of years. So when God said, "Let there be light," the process was probably already going on, and took thousands to millions of years. Some of the creation days in the Bible probably took longer than others.

Wow! What a revelation. This is a new way of thinking. The explanation of the evening and the morning, as described at the end of each " creation day," was really meant to set the stage for the future generations of Israelites to adhere to the weekly Sabbath. After all, God didn't need to rest on the seventh day, but it was compulsory for the Israelites to do so, or to face death.

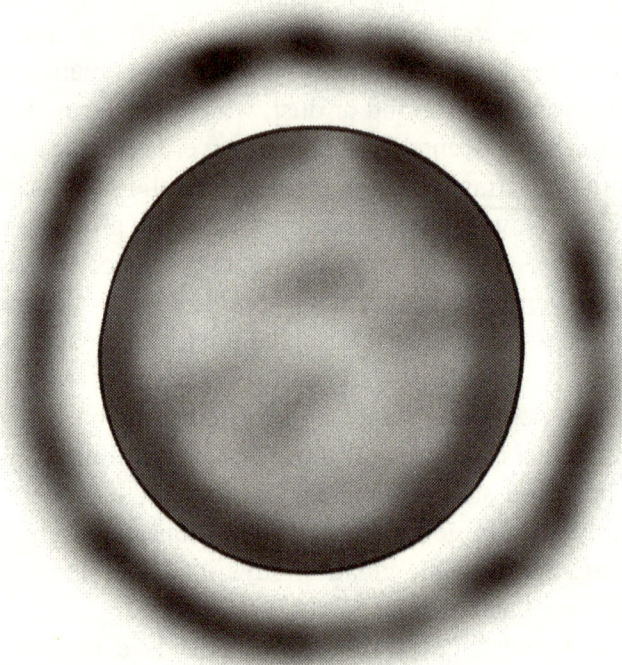

Figure 7-Picture of the earth covered by a thick
cloud which sunlight cannot penetrate

What was the purpose for the light in <u>verse 3</u>?
According to evolutionists, the first abundant fossils of living
organisms, primarily bacteria and single cell structures,
occurred in the Proterozoic Era, some 2.5 billion to 550
million years ago. Many of these organisms were early
microcellular living cells and organisms that produced or
manufactured oxygen. These were very primitive and simple
cell structures, but they played a significant role in evolution.
By the middle of the Proterozoic Era, we see evidence of an
increasing concentration of oxygen in the earth's atmosphere.
These organisms, therefore, used light as a power source

to produce oxygen. This obviously took millions of years. This oxygen atmosphere changed the world forever. Without oxygen, the earth would be a very different place.

As the content of the atmosphere changed to oxygen predominance, some organisms were wiped out of existence or made extinct, while others survived (survival of the fittest). Thus, this opened the door for the evolution, God-directed, of other life forms that were pleasing to God, or that were necessary at that time to accomplish His purposes. These life forms were necessary at that particular time for God's purpose, but that didn't necessarily mean that they would survive later. Light was the source of energy for these prehistoric organisms, as it is today for many life forms. We see the same thing happening today in simple algae and complicated plant species. Well, you're probably saying that I am now assuming that evolution is occurring, but how do I know that God is directing "creation?"

Let's jump ahead:

Gen 1: 24: And God said,
Let the earth bring forth the
living creature after its kind,
cattle, and creeping thing,
and beast of the earth after
his kind; and it was so.

Contrast this with

Gen 2: 8: And the Lord God
planted a garden Eastward in

Eden; and there He put the
man whom He had formed.

And

<u>Gen 1: 26</u>: And God said, Let
us make man in our image,
after our likeness...

And

<u>Gen 2: 7</u>: God formed man
out of the dust of the ground,
and breathed into his nostrils
the breath of life; and man
became a living soul.

Notice the difference between the first verse and the other three: <u>the earth was allowed to produce</u> living creatures. God was directing evolution, while in the other verses, He <u>physically planted</u> the Garden of Eden. He specifically said that they are going to make man whom he shaped out of the dirt. And He breathed into his nostrils the breath of life. Notice this secret and big difference. Therefore, God was allowing the process of evolution to occur; this could only have occurred with and by the direction of God. <u>Therefore, despite the fact that evolution is statistically impossible, just by the odds against it, it became a reality by God's intervention</u>. He physically made man; the earth did not bring forth the human race. Also keep in mind for later that at the end of creation/evolution, the earth had to be hospitable for mankind.

So even before the fish and birds were created in their present form, <u>CREVOLUTION</u> (creation/evolution: God-guided creation in evolution) was already ongoing and had a divine purpose. The ancestors of fish and birds were crevolving millions of years before they became what we have today. The ancestors of all living things that we can see today were already crevolving millions of years ago. The sea and land had to be able to support life to come, as well as be able to – bring forth new species of life. Light and oxygen were made available to sustain life that was exploding in number and variety during the Proterozoic Era.

Now let's continue on with Genesis.

<u>Gen 1: 9</u>: And God said, let
the waters under the heaven
be gathered together unto one
place, and let the dry land
appear; and it was so.

Remember that the earth was completely covered with water, and the waters were teeming with early life forms. As a reminder, these life forms were the precursors to all life that currently exists, and has ever existed on the earth. Now come violent earthquakes to raise the land above water and create deeper oceanic valleys to make the dry land appear. Have you ever experienced a really strong earthquake? Now sit down and imagine what strong earthquakes would be needed to raise the land masses we have today. Won't you see the fossils of those organisms all over those raised land masses? That's exactly what scientists have discovered. Think of the enormous tsunamis that were occurring.

Think about what the continents were doing at the time, moving and separating out the way we have them today. In the Cambrian period, 544 million to 500 million years ago, the continents were joined in one large, single continent, as you learned in geography. This large, single mass of dry land was called Rodinia. The dry land appeared. This rather large mass fragmented and migrated to the present locations that we can see today on a world map that shows the continents and land masses. This took millions to billions of years. Everything that God was designing was taking millions to billions of years. To us, it may seem an eternity, but to God, it's nothing.

In the meantime, going back to the Proterozoic Era, 2.5 billion to 550 million years ago, we see the fossil remains of soft-bodied organisms, similar to what I'm sure you've seen in museums. You can see the imprints of these creatures embedded in rock. Most scientists postulate that many other simpler life forms probably existed prior to this, but they did not fossilize, and thus we have no record of them. However, we still see the slow emergence, or evolution, of more complicated life forms that the earth was bringing forth. Again, this was directed by God.

<u>Gen 1: 11</u>: And God said, Let
the earth bring forth grass,
the herb yielding seed, and
the fruit tree yielding fruit
after his kind, whose seed is
in itself, upon the earth; and
it was so.

If one looks at the appearance of fruit trees with seed by examining the fossil record, we have to jump into the future to the Devonian period, approximately 400 million to 350 million years ago, which is millions of years ahead. That is when we see the emergence of all sorts of trees, ferns, and their relatives. Therefore, when God gave the order for these trees to be formed, and it was so, the writer was giving us a summary of the event. (Whether the author or authors knew about the time span of creation being millions to billions of years is not sure. I seriously doubt that he or any one else, meaning the biblical authors, knew this information). In reality, this event, like other events in creation, took millions of years, and the process of evolution was going on millions of years before a particular organism was "created."

I cannot emphasize more the process of an ongoing evolution of living things described by scientists and corroborated by the Bible. Evolutionists describe their findings in the strata of the earth as one of an early beginning with simple, one-celled organisms, to a slow evolution, to very complicated living structures. The Holy Bible is not in disagreement with that statement.

Figure 8-Picture of a prehistoric forest

The Paleozoic Era is famous for the explosion of numerous life forms, including animals, fungi, and plants. The land was slowly being invaded by these life forms as a new ecological frontier, and it is also believed that insects became airborne during this period. Overall, we see a slow crevolution of organisms over the millennia. Now remember God hadn't gotten to the "fish and birds" yet, although their precursors were already evolving. So the land was being prepared for the coming of the birds, and the fish and other marine life were being "developed" to coincide with the age of the birds.

We see the complex relationships amongst the fish, birds and plants. In fact, the Ordovician period, which occurred millions of years before the Devonian period, is best known not only for its diverse marine invertebrates, but also for primitive land plants, suggesting that plants probably invaded the land during that particular period of time. Plants then evolved to become complex structures throughout the Ordovician, Silurian and Devonian periods, and continue to do so with much more complexity as they continue to evolve during the following millions of years. It's in the Devonian period, 400 million to 350 million years ago, that one sees fossils of ferns, horsetails, and seed plants.

<u>Gen. 1: 20</u>: And God said,
Let the waters bring forth
abundantly the moving
creature that hath life, and
fowl that may fly above the
earth in the open firmament
of heaven.

47

This process was going on for millions of years when this "order" was given. Again, the writer was giving us a brief, simplistic summary of creation. The writer probably didn't know about the complexity of creation and the millions of years it took, but envisioned it the way he would look at the seven-day week system. On the other hand, God, through making it simple for man to understand, surely knew exactly what He was having the writer record. The Silurian period, 450 million to 400 million years ago, is well known for the evolution of fish. Though "simple," these later evolved into complex fish. Don't forget the numerous other moving creatures that, in themselves, were embarking on their own evolution. On land were the ancestors of spiders and centipedes, and the earliest fossils of vascular plants.

<u>Gen 1: 20:</u> …and fowl that
may fly above the earth
in the open firmament of
heaven.

The fossil record of birds leaves a lot to be desired. The scientific community hasn't really found that cache of avian fossils they are dying to have. It is believed that the thin, light, hollow bones of birds were not amenable to the process of fossilization. The oldest known fossil identified as a bird is still the dinosaur-like Archaeopteryx of the Jurassic period, which we have all seen in museums and books. It is believed that birds evolved from the reptiles, and, in fact, there is evidence of reptile-like creatures with wings. There were precursors of other birds at the time, however, but we have not located them yet.

If you look at the slow evolution of every other species, it is natural to assume that the birds underwent the same process. Just because we cannot find their fossilized ancestors (yet), or even if they don't exist (meaning they are forever lost because of decay), that doesn't mean that they didn't evolve like everything else. Birds continued to evolve in variety and function throughout the Cenozoic Era into what we have today. So even though the marine creatures and birds were created on "day" five, you must again keep in mind that this creation day took millions of years, while other creatures, plants and animals were undergoing their simultaneous crevolution. A creation day was not 24 hours or 1,000 years.

<u>Gen 1: 24</u>: And God said,
Let the earth bring forth the
living creature after his kind,
cattle, and creeping thing,
and beast of the earth after
his kind; and it was so.

Figure 9-Picture of numerous animal herds

We know cattle didn't appear on the earth right away. As we have seen, everything underwent its own evolution, so it's natural to assume that cattle did the same. Let's look at the evolution of the mammals. The Cenozoic Era, approximately 65 million years ago to the present, is known as the age of the mammals. However, it can also be called the age of the insects, flowering plants, fish and birds. This just proves to us that the process of crevolution was an ongoing process from the very beginning, and had been ongoing for millions of years. Getting back to the mammals, their history began millions of years before the Cenozoic Era. One of the most unmistakable evolutionary landmarks in the Carboniferous Era, 350 million to 285 million years ago, was the amniotic egg. This allowed the ancestors of mammals, as well as the ancestors of birds and reptiles, to reproduce successfully on land. This was an important evolutionary step for mammals, but it by no means occurred suddenly. It is through the Mesozoic Era, 245 million to 65 million years ago, that we find the first fossils of modern mammals. Their ancestors were obviously being crevolved millions of years prior to this.

CHAPTER 9
THE CREATION OF MAN

Figure 10-Picture of God creating Eve with Adam in the
background

Gen 1: 26: And God said, Let
us make man
IN OUR OWN IMAGE,
AFTER OUR LIKENESS.

And

Gen 2: 7: And the Lord God
formed man out of the dust
of the ground, and breathed
into his nostrils the breath

of life; and man became a
living soul.

Can you see and understand the similarity between these two verses and the difference with the ones prior? <u>Here, God is directly and physically involved in the creating process</u>. The earth wasn't allowed to bring forth man; rather, God made man "with His own hands," and directly gave him the breath of life. This brings up many questions regarding prehistoric man, which I will attempt to answer later on. <u>God made man!</u> He was not a by-product of evolution, but rather God-made. I'm not finished yet! Some of you might be thinking that everything was going okay until I came to the creation of man. For a while, I was stuck and confused on this very same issue until I came across some hidden secrets. Why am I, the author, neglecting the "evolutionary evidence" of man? I will get to this later. I have intentionally made this chapter short so that it will stand out Maybe you should read it again before you proceed.

CHAPTER 10
<u>WHAT'S GOING ON?</u>

So, in summary of the above, I believe that I have been able to link both the "theory of evolution" and the "story of creation" with respect to non-human living creatures: the fish, birds, animals and plants. The creation days, or time periods of the days of Genesis mentioned in the Holy Bible, were probably millions to billions of years in duration. Plants and animals weren't made on a daily sequence as we think it is portrayed in the Bible. It was not meant to be taken literally. Rather, it was God's simplistic way of providing man with a synopsis of the events that occurred and linking it with the weekly ritual of the Sabbath. <u>In fact, the crevolutionary processes were all overlapping, yet the appearance of living organisms more or less matches the daily</u> creation sequences <u>seen in Genesis. Again, this sequence that the Bible discusses matches what man</u> has discovered. When God gave the command to create, for example, the birds, I believe He was picking and choosing life forms that were suitable to His plan in the future; that plan included a fairly hospitable earthly environment where the human race was to live and thrive. Remember that what the earth was bringing forth was life in great diversity, complexity, and wonder. But this was initiated and directed by God Himself.

Let's look at, for example, the modern-day telephone. Look at what the telephone looked like and how it functioned years ago at its invention. Let's look at it now. That's technology in evolution. Imagine also different styles of telephones being invented at the same time, but only a few becoming popular with consumers as time went on.

The companies making the unpopular telephones would go out of business, and those telephones would become "extinct." Other companies survived (survival of the fittest), and continued to produce the more successful complex telephones that science had to offer.

That is what crevolution is all about, and it is directed by God who is the "consumer" in this case. During the crevolutionary millennia, thousands of species were wiped out or made extinct. There are numerous explanations for these extinctions, from temperature changes, to meteors, to the Ice Age, to the predominance of other species. This I shall leave to the scientists. That includes the extinction of the dinosaurs and prehistoric human-like forms, the latter of which were wondrous products of the earth "bringing forth" its many life forms. They were all brought forth or made by the earth, not directly by God, but indirectly. The Holy Bible has proven itself when it comes to actually proving how and when evolution (crevolution) actually occurred. Is it so hard to believe that modern-day man and "prehistoric" man were two different species? We'll soon find out.

I would like to briefly take a tangent off course to remind you about the apparent contradiction in the Bible with respect to the creation "stories" described in the two chapters of Genesis, narrated by the so-called Priestly and Yahwistic Traditions. The Priestly account reflects the ritualistic traditions of the Jewish religion, and one can read their account of creation from <u>Genesis 1:1 - 2:4</u> (first part of verse four). The Yahwistic account of creation can be seen from <u>Genesis 2:4</u> (the last part) onward, and reflects on the role of God, Yahweh, and His interaction with His chosen people, Israel, as well as the rest of mankind. There is a lot of argument amongst theologians and biblical scholars as to the exact date of these two accounts, and they believe that these accounts may be written by a different author or a different group of authors. In fact, these two accounts are the same and do not contradict each other, as you will soon learn below (see Chapter 6). They have their opinion, and I'm entitled to mine.

As you learned in the preceding pages, there is a big difference between "the earth bringing forth" and God directly creating something. Genesis 2 implies that Adam was created prior to the animals (Gen 2:7-20), which contradicts the first chapter of Genesis. The hidden truth is that there is no contradiction here! Genesis 2:19 says that God (directly) formed the animals out of the soil. These animals were animals specifically created to live in the Garden of Eden and would have been more domesticated and friendly than those living out in the wild. Remember that the animals outside the Garden of Eden were the products of millions and millions of years of crevolution. It's therefore possible that the animals created in the Garden of Eden were totally different from those that crevolved, or were made to keep that ecosystem sound. He could have made some or all of the domesticated animals in the Garden of Eden. That is not to say that all of the animals outside the garden were a threat to Adam. On the contrary, the animals weren't a threat to mankind. They were to co-exist with man There were lots of other animals living outside the garden which the "earth brought forth." God specifically created those animals to co-exist with humanity, but not to live in the Garden of Eden, which He planted (Gen 2: 8, 9). Some creatures could have trampled the Garden of Eden in no time if they were to live there, making the work of Adam and Eve a huge burden. The possibilities are endless; choose one.

Now let's get back to man and the so-called ancestors of man. First, let's look at where we are now. Ignoring social issues, war, conflicts, etc., man has seemed to live, colonize, and explore most of the earth's diverse environments, as well as space. He has advanced technologically to help him live well in these environments, conquer a lot of diseases, and produce enough food to feed the billions of his offspring. We have gone to the moon and are on the verge of exploring the vast unknown of space. His average life span is in the mid-70's and is expected to increase, provided he doesn't follow an evolutionary path of extinction. Where did we all come from? Did we evolve or crevolve? Were we created, or did we evolve over millions of years? Did God make an exception of us and instantly create us in "His own image and likeness?" Let's explore and analyze both opposing views.

CHAPTER 11
<u>EVOLUTIONISTS ON MAN</u>

Most people have heard that humanity evolved from the apes. In fact, according to the evolutionists, the ancestors of present-day apes, chimpanzees, and humans diverged as separate species from a common ancestor about five million years ago. Collectively, we (apes and humans) are called Humanoids. The human species, from the common humanoid ancestor to present-day man, is called Hominids, while the ape species from the same humanoid ancestor is called Pongids. Gorillas are thought to have diverged from a much earlier hominoid, approximately eight million years ago.

The science of human paleontology is plagued with a lot of guesswork based on discovered "human" skeletons. I believe that because of their belief in evolution of all life forms on earth, these scientists try to link their beliefs in human evolution with the discoveries. Nevertheless, most scientists believe that, as far as human evolution is concerned, Australopithecus afarencis evolved into Australopithecus africanus, which evolved into Homo habilis, Homo ergaster and Homo erectus. Here, a great debate is ongoing, whether H.erectus or H. ergaster was the ancestor of modern humans. From one of the two species previously mentioned, Homo heidelbergensis, previously referred to as archaic Homo sapiens, evolved. This species then led to the evolution of both the Neandertals and modern day humans (Homo sapiens sapiens) as two separate species. Many of these species co-existed and interbred with one another.

The first hominid thought to have appeared was the Australopithecus species, a bipedal ape (some of the time); one of which, notably A. afarensis, is thought to be the direct ancestor of Homo habilis about two to three million years ago. All Australopithecus species became extinct approximately one million years ago. A. africanus, which evolved from A. afarencis, lived in Africa, was approximately 1.5 meters tall, weighed about 220 pounds, and had a small brain (450 c.c. approx. vs. a human brain > 855c.c.). This ape-like creature is not thought to have used any significant tools, and definitely didn't make use of fire. It is also felt to have been heavily built for its height, and it climbed and swung on trees. In general, it was a less active creature than its descendants. There is some recent thinking that the male and females of the species A. afarensis showed fewer differences in size than earlier thought. It is now felt by some that they cooperated with each other, male and female, and therefore showed early signs of human behavior. Keep in mind that some of these issues are still very controversial and may even be conjecture.

Homo habilis, or the " handyman," as it is sometimes referred to, appeared on the scene about two million years ago, and, like A. africanus, lived in Africa. This species is considered by many to be the first humans. It was first discovered in 1959 in Tanzania. In addition, a nearly complete skull was discovered in Koobi Fora at Lake Turkana in Kenya soon afterwards. It is distinguished from the former in that it was a maker and user of tools and had a larger brain (700 c.c.). Tools included rocks and sticks, mostly very primitive in nature. It was also ape-like and bipedal, like its predecessor. It was approximately 1.5 m tall and weighed about 100 pounds. It had a similar face to

Australopithecus, but the nasal bones were more protruding. H.habilis could have walked upright, and, like A. africanus, could climb trees. Some even speculate that it had at least the beginnings of speech and was able to identify and solve problems. Examinations of H. habilis skulls reveal that they were mostly right-handed. There is even evidence suggesting that there was a division of labor between males and females. The males were hunters, and the females were gatherers and probably roamed in large groups.

Homo erectus appeared in Africa about 1.5 to two million years ago, and they also lived in Asia, particularly China. The earliest discoveries of Homo erectus were in the Rift Valley of Africa and South Africa. It co-existed with A. afarensis and H habilis. These three species were morphologically similar, yet they are considered as separate species. H.erectus stood approximately 5.5 feet tall, weighed about 100 pounds, and had a larger brain volume of about 750 c.c. to 1300c.c. Males were much larger than the females. They were more efficient at tool making and used things like cleavers and hand axes, as compared to Homo habilis. Some postulate that the design in the tools of Homo erectus showed intelligent reasoning and advanced thinking and capabilities. These tools required greater hand-eye coordination and skills, which, apparently, Homo habilis had not quite developed yet. They regularly made use of fire and were successful hunters. Many believe that Homo erectus was probably the first "human" species to migrate out of Africa into the European and Asian theatres. It was also highly likely that this quest out of Africa was made successful because of the use of fire, coupled with good tool design. Many believe that H. erectus was not the direct ancestor of H. sapiens sapiens (modern man), but the intermediary Homo heidelbergensis, formerly "archaic H. sapiens," was that link. Yet others believe that Homo erectus was the missing link, and therefore the direct ancestor that led to modern humans. One must realize that a lot of these theories and conclusions are all based on conjecture.

Homo ergaster roamed the earth about two million years ago throughout Northern and Eastern Africa. One of the prominent discoveries was found in Lake Turkana in northern Kenya. It was about 1.6 meters high and weighed approximately 140 pounds. Its skull was like a football in shape, and it had a brain capacity of 850 ccs. There is evidence that they had primitive tools and had the beginnings of social and behavioral traits consistent with humans. Some believe that they were the first to migrate out of Africa into Asia and Europe, rather than Homo erectus.

What about the Neandertals (formerly Neanderthals)? Some think that the Neandertals were not humans and branched off in a different evolutionary direction from "archaic Homo sapiens" or Homo heidelbergensis. This divergence occurred well before the appearance of modern humans. It was archaic H. sapiens that later "evolved" into modern humans. They (the Neandertals) were approximately 1.6 meters tall, and weighed about 175 pounds. They had populated the European continent from about 30,000 to 150,000 years ago, after which they became extinct. They had lived in ice age conditions and were forced to compete in a very dangerous environment. Neandertals were excellent hunters who used a variety of tools in hunting and gathering. They didn't use bows and arrows or projectiles. Their speech areas were not as developed as modern humans. They were heavily built, having very thick arms and leg bones, indicating tremendous strength. They had coarse facial features and large brains (1500 c.c. average). Their skull and body features were very different from archaic H. sapiens. They even buried their dead. They are now believed to have been much smarter than previously thought.

Homo heidelbergensis, formerly archaic Homo sapiens, thrived about 800,000 years ago. This classification encompasses a diverse group of skulls that share many features with H.erectus, the Neandertals and modern humans. There are very few fossils that have been discovered so far. They were about 1.7 meters tall and weighed approximately 130 pounds. They also had an increased brain capacity of 1600 ccs.

Modern man, or Homo sapiens sapiens, flourished about 35,000 years ago, approximately the same period the Neandertals became extinct. Some fossil remains date from 60,000 to 80,000 years. Most, if not all, of the so-called cave drawings were believed to have been drawn by the earliest "form" of Homo sapiens sapiens. He progressed to the civilization that we see today. I underlined the word "about" above because the exact time period may never be known.

Recently, based on certain evidence, some paleontologists believe that instead of having one "out of Africa" migration to Europe and Asia, there could have been three migrations occurring at different times. Some believe that these three waves of immigration occurred about 1.8 million years ago, about 400,000 to 800,000 years ago, and, finally, about 80,000 to 150,000 years ago. So instead of Homo erectus or Homo ergaster moving into and populating the European and Asian arenas in one move, and later evolving into modern humans, some paleontologists now believe that modern day humans immigrated out of Africa about 100,000 years ago into the playgrounds of Homo erectus and the

Neandertals. As a result, there was interbreeding among those species to form what modern humans look like today.

There certainly is a lot of confusion and speculation on this matter. In addition, could man's dating methods be slightly off? And if so, by how much? I don't know, but we need to think about it. However, what I would like to draw your attention to is what I have suggested later in Chapter 13—that the descendants of Adam were around and roaming the earth long before the pair sinned in the Garden of Eden. For how long I don't know, because the Bible doesn't give us any clues, but it could have been for thousands of years. Remember this paragraph when you get to Chapter 13.

Figure 11-Picture of a busy downtown city

What does the science of genetics add to this evolutionary scenario? Well, let's look at some of the genetic evidence, even though the experts themselves admit that a lot of work still has to be done and, at present, there is so much confusion. Today, a person's innocence or guilt is stronger, based on genetic DNA testing, as compared to eyewitness testimony. I believe that accurate genetic investigation would provide us with a better "evolutionary theory of man," or refute it. DNA (deoxyribonucleic acid) analysis, primarily of mt. DNA and Y-chromosomes (see definitions), indicate that we, as humans, Africans and non-Africans, all have the same ancestral home of Africa. Whether we like it or not, we all came from that part of the world. Our ancestors emigrated all over the world from that locality. To emphasize, that general vicinity was the origin of modern man, as recorded in the Bible. The exact migration date is unknown, but it is believed to have occurred approximately 80,000 to 150,000 years ago. Again, modern humans, and I don't mean that in the evolutionary context but the Biblical one, have a common "African" heritage. I know that this paragraph is expressed in a simple fashion, and that the search for genetic information to decipher our past is presently ongoing and complex. I must admit that I do not have all of the information, nor do I fully understand all the genetic evidence available.

Some experts divide humanity into four groups: a) African b) Caucasians c) Mongoloids and d) Australian. Africans apparently have the most genetic differences from the others, indicating that Africans split, and thus changed very early in the family tree. Australian Aborigines are most distant from the Africans based on genetic studies. The last common ancestor for all non-African humans is thought to have existed about 50,000 years ago. Therefore, everyone outside of Africa came from the same small band or group of humans that migrated worldwide approximately 80,000 to 150,000 years ago. Every human being, African and non-African, all came from one common ancestor. "In fact, in India, all of the early lines that gave rise to Asians and Europeans are found in great profusion and antiquity," said a DNA tracker, Stephen Oppenheimer of Oxford University, in a press teleconference.

Could so much of evolution in race, color, features, religion, language, etc., have occurred in just 80,000 to 100,000 years, despite the odds, while it took millions of years for other species to crevolve? Is evolution in humans different from that in animals? Are the odds different for human evolution than for non-human evolution? I don't think so. Why did man encounter a more rapid evolution than everything that I discussed earlier? It is also during that time we developed language, which is considered unique to humans. That not only includes speech, but the drawings that we see in caves all over the world, dating approximately 35,000 years ago. I would say that that was pretty fast. Think about it! It just doesn't make any sense. I would like the paleontologists to explain that phenomenon. Something is missing here.

CHAPTER 12
<u>CREATIONISTS ON MAN</u>

What do creationists believe? What do they think they believe? God created man instantly on the sixth day of creation, just like He created everything else. Some believe that the creation days were 1,000 years long. Does that mean God took such a long time to make the different species? I don't think that they have an answer to that question. Nevertheless, He formed Adam out of dirt like a potter forms a pot out of clay, and blew the breath of life into his nostrils. Man then became a living soul. I wonder how long these people think that it took God to form man out of dirt during that 1,000-year period? He then planted a garden, the Garden of Eden, for man to live in and to guard; but guard from what? He then created Eve from Adam's rib and blessed them, telling them to multiply and fill the earth. However, before they could settle down in the Garden of Eden and have a family, Satan came to them in the likeness of a snake to discredit God and to tempt Eve. As the narrative goes, they both sinned and disobeyed direct orders from God, and were then kicked out of the Garden of Eden and doomed "forever." It was after their eviction from the Garden of Eden that the pair first had intercourse and had Cain and Abel. Man multiplied in the Middle East/African area until his population was cut to only Noah and his family by the worldwide deluge, the Flood. Later, God confused their language, and man was dispersed all over the globe. They eventually looked different, spoke different languages, and developed different cultures (Remember that all of humanity possesses 99.9 percent similar genes). From Adam until now, man has been in existence for approximately 6,000 to 12,000 years, depending on whom you talk to. It is that simple and straightforward. What a different scenario compared to evolution. As you will see later, not all of this is correct.

The whole basis of the belief in creation rests on the Holy Bible. But we really don't know the Bible. We read it over and over but do not comprehend it. All we know is what we were taught since childhood, and what is ingrained in our minds. A lot of us stand firm with the teachings of the Bible, but have not researched the issues to prove and strengthen that faith. We have not "proved all things." We all follow religious leaders who do not know and understand the Bible. That is what the Bible refers to as the blind leading the blind. It's either a book from God, or a book written by numerous authors and then twisted by our religious leaders to deceive the public. These religious leaders make us believe that one has to have unequivocal and total faith in this book, the Bible, a "blind" faith as it were, and that one has to accept this belief without any concrete evidence to support any sort of argument. What an awe-inspiring faith that can be. If you believe in the Bible, then you take the stance of creation. If you do not believe in the Bible, then you believe in evolution and thereby appear atheistic. There is no middle ground. You are either hot or cold, as it were. Is it possible that both creationists and evolutionists, because of a lack of biblical wisdom and knowledge, are interpreting the Bible differently, and yet the basics are the same? I think that one group looks at a glass half filled with water as half full, while the other looks at it as half empty.

People create their own gods to suit their own individual needs, or they might take God out of it altogether to accomplish the same. Or one might be an evolutionist and believe in God, but on one's terms only. Look how many of us believe in the Bible, but not the Old Testament. Yet others believe in Satan and not God, claiming that God is dead. The creation survives the creator. Isn't that ridiculous? There is so much confusion. No wonder there is a large rift between evolutionists and creationists, and this obviously has been the topic of numerous heated debates. In this book, I have showed you the theory of crevolution with respect to plants and animals, thus confirming the authenticity of the Bible and thereby linking creation and evolution. On this topic, the Old Testament has proven itself.

What about the timeline issue? Is that important? What's important to know is that the creation days took millions and millions of years, and not one day or 1,000 years. So what if the scientists are off a few hundred thousand years? Some are held hostage by their methods of determining the age of geological finds, which may or may not be that accurate. As a result, they are subject to a lot of ridicule by the creation purists. I, myself, am not an expert, and I simply do not know. In fact, it really doesn't matter that much to me because everything fits like a glove on one hand with respect to the crevolution of plants and animals, as you saw earlier. On the other hand, the other glove fits perfectly with respect to the creation of man, as you will see in the coming chapters.

CHAPTER 13
<u>WHAT THE REAL TRUTH IS!!!</u>

What is the real truth? How does the Bible prove itself again? The Bible was proven true and reliable when we discussed crevolution of plants and animals earlier; can it do so again when it comes to the creation or evolution of man? You bet, and we shall soon find out.

First of all, man didn't evolve or crevolve. He was created! The earth didn't bring forth Adam. God formed him out of the dirt and blew the breath of life into his nostrils (<u>Gen 2: 7</u>). This was a different and specific action from God. When he made woman, he took a rib from Adam (which contained his genetic DNA information in the bone marrow), and that's why Adam said, "...bone taken from my bone, and flesh from my flesh" (<u>Gen 2: 23</u>). Woman was also made by a direct and specific action by God. Some people think that men are missing one rib, which would definitely corroborate the creation story; that is not true. They both had the same genetic information, a clone, as it were, except Eve was female. God used the same genetic information to form Eve but altered the chromosomes to form Eve. She was a genetically altered clone of Adam. After He created Adam and Eve, He told them to be fruitful and multiply and fill the whole earth (<u>Gen 1:28</u>). <u>Hold this thought, and remember they were told this before they were kicked out of the Garden of Eden.</u>

Now we come to <u>Gen 3:1</u>: "Now the snake was the most...etc." You have to throw away your old thinking, the one you grew up with, and open your mind to a new way of thinking. What was the time period between the creation of

Adam, and the creation of Eve? Is the time interval between Gen 2:25 and Gen 3:1 a period of one day, one week, or more? Could it have been hundreds to thousands of years? When Cain was banished after he killed his brother, Abel, he was already married. Therefore, Abel and Cain weren't the first two children of Adam and Eve, as some would have you believe. The Bible does not state that Cain and Abel were the first two children of Adam and Eve. Other children and descendants were not mentioned or omitted for certain reasons. Also, some experts insist that the murder of Abel was the first murder ever committed, and that murder was accidental. Cain didn't know how to intentionally kill a human being and probably never saw a dead human. Come on! I totally disagree with this interpretation of the Scriptures! God put a mark on Cain's forehead so that others wouldn't kill him. Where did these "others" come from?

There have had to have been previous murders prior to that of Abel, considering the large number of people living in the region at the time, many of whom had turned their backs on God. One of the reasons that this story is mentioned is to show the reader how Seth, a future son of Adam and Eve, became part of the lineage of Jesus Christ. As one can see, this lineage was mentioned several times in the Old and New Testaments. Seth took the place of his slain brother Abel to become one of the forefathers of Christ.

After the pair sinned, God told Eve in Gen 3: 16 that He would greatly increase her trouble in pregnancy and childbirth. That highly suggests that she may have had many, or some, "painless" childbirths in the past. There must have been, therefore, a very large population of people living in the area and the surrounding region, or even worldwide, for "them" not to know Cain. That is why Cain had a mark on his forehead; whatever that mark was, I have no idea. So Eve had many, many children before the pair sinned against God. So while the pair enjoyed the blessings of the Garden of Eden, their descendants were busy colonizing the world. It is also possible that the only true legal residents of the Garden of Eden, other than specific animals, were Adam and Eve only. I do not mean to assume that there were millions of people alive at the time, but maybe several thousands, or possibly even hundreds of thousands. Also, if God wanted to kill all of humanity, why did He have to flood the whole world if they were all located in the Middle East/ African region? We have seen in different parts of the Bible where He was effectively able to take out a local town or population. Why bother Noah and his family to construct a huge boat that would take over 100 years when it wasn't necessary? Why did He bother to save certain animals in Noah's Ark, if there were others alive elsewhere, and if He wasn't planning to destroy life as we know it on the whole earth?

I believe that there were many people, all of whom were descendants of Adam and Eve, who, according to Genesis 1: 28 were starting to populate the whole earth. And this was going on for hundreds to possibly thousands of years before the temptation of Adam and Eve in the Garden of Eden. This would definitely explain the discovery of "modern humans," who were discovered several thousand years ago, a discovery which cannot be explained by creationists, based on a human existence of 6,000 years. But one might say that Adam was 130 years old when he had Seth (Gen 5:3), and therefore Abel and Cain were born when Adam was much younger. How does that equal to thousands of years? If I cannot explain this contradiction, then my whole theory is out the window. I believe that Adam's age began to be counted right after he was kicked out of the Garden of Eden because he knew that he was going to die. Prior to that, he was supposed to live forever, so what's the point of counting one's age? I bet that Adam explored the earth years before Eve appeared. Adam himself may have drawn some of those figures that we see in some caves today. Adam could have been thousands of years older than Eve. After the pair sinned, their days were numbered.

Only one tree that "gave life" was planted in the Garden of Eden. When they were kicked out of Eden, they weren't allowed access to the tree, or else they would live forever. Therefore, it's logical to assume that, as the descendants of Adam and Eve multiplied and filled the earth, they would have had to make a pilgrimage to Eden every few hundred years or so to eat from the tree of life to be able to continue to live forever. After all, Adam lived for 930 years after he was kicked out of the garden. One can ask how an earth filled with people could eat from one tree. Can one tree bear enough fruit for possibly millions of people? With God, everything is possible, but you must remember that it was probably not God's thought at that time to accomplish such a plan. I wonder if, and how many, people died and became fossils because they weren't able to return to Eden and eat from the tree of life while Adam and Eve still lived there.

The people who traveled the earth over these thousands of years had to bring the earth under their control (Gen 1: 28). They had time to develop amazing technologies, which we, today, cannot duplicate. I wonder if the technology to travel the oceans, or to build the pyramids of Egypt and South America, as well as the wonder of Stonehenge, could have had its origin in that time. I wonder how much technology was lost at the time of the flood. And remember, these descendents of Adam and Eve could have had some help from God directly or indirectly through angels.

Now the obvious questions that come up are why did the rest of Adam's descendants have to be condemned just because two people, Adam and Eve, sinned? Why was such a burden placed on Adam and Eve? Why didn't God punish them both in whatever manner He wanted and leave their descendants alive to fill the earth? Why did all of humanity have to suffer, even until today, just because Adam and Eve sinned? Why did the actions of Adam and Eve affect us all? Why are we paying the price of their error? Why was that particular sin so critical? Did the descendants of Adam and Eve commit sins because, like Adam and Eve, they were susceptible to temptation? Did those sins count? Were the sins of the people considered insignificant because, in reality, all their sins were minor, and, simply put, humans were made imperfect? Because of one man, sin entered into the world and condemned the human race. These and other questions will be answered at another time.

Let's talk about the ill-fated city of Atlantis, for argument's sake. I cannot verify whether it existed or not. Those who think that it did exist at one time cite past scholars centuries ago who made reference to it being destroyed 9,000 to 12,000 years ago by a catastrophic flood. They believed it was buried somewhere, possibly in the Atlantic Ocean. By the way, those same people ignore the "story" of a worldwide flood, common to many cultures, which refers to a flood and Noah. Could a city like the city of Atlantis exist in the past?

I would say to that a definite YES! In fact, I believe that several cities similar to Atlantis existed, all of which were inhabited by the descendents of Adam and Eve. Initially, they were a God-fearing people, but when Adam and Eve were kicked out of the Garden, they spiraled down to where they were wicked, and even their thoughts were evil all of the time <u>Gen 6: 5.</u> Prior to Adam and Eve's disobedience to God, the city of Atlantis, as well as others, were thriving centers all over the world. Now comes the flood, which destroys them all. It didn't take a very long time for the flood waters to recede, so, most likely, deep valleys and massive refacing of the earth occurred by underwater earthquakes. Those cities are buried under water perhaps, which used to be dry land prior to the flood. Just another thought—the vast distances between the African/European continents and the Americas may not have been all that far apart.

Another evidence of a worldwide dissemination can be seen by comparing fossil evidence with Gen 6:18. Giant footprints were found in the Americas alongside "modern human" footprints. I haven't found much on this particular discovery, but I find it very interesting. This is far from the area in which Noah used to live. These giants lived on the earth before the flood. Therefore, people used to live in the Americas at the time of Noah. However, only Noah and his family were found worthy of living at the time of the flood. God had to deluge the whole world in order to kill everything.

One more piece of evidence emerges. "Modern human" footprints and handprints were found in the same area as some dinosaur footprints. Were some species of dinosaurs alive during the time of these "modern humans," and were they eventually destroyed by the flood? These so-called "modern humans" were the descendents of Adam and Eve who were exploring and trying to control the earth for thousands of years while Adam and Eve still enjoyed life in the Garden of Eden.

So what happened after the flood? According to <u>Gen 11:1</u>, at that time, the people of the whole world had only one language. The "whole world," according to <u>Gen 10</u>, included parts of Africa and the Middle East, as well as parts of Europe, including Spain. Nothing is mentioned about the New World. These were the new inhabitants of the earth, all direct descendants of Noah (<u>Gen 10: 32</u>). Remember that these people still had the knowledge of their ancestors, who traveled across the world at the time of Adam and Eve. They had the know-how to build the tower of Babylon, which was supposed to "reach the sky." This was obviously not a two-story tower that you might be thinking about. They wanted to build the tower so that they wouldn't be scattered all over the earth, as their ancestors had been. They obviously had the technology to do so, as well as information about the earth and how vast it was.

Getting back to the tower, even the Lord was impressed with the undertaking, as we read in <u>Gen 11: 5-9</u>. We read that He was concerned that these people would soon be able to do "anything they wanted to do." That came from the Lord himself. They therefore had to have possessed the information to create amazing technology, for their time period, and in the near future.

I believe that if God had not interrupted mankind at that time, we, as a human race, could have developed cures for many of the diseases that plague us today. In addition, our inventions, like the airplane, communication devices, and others would most likely be hundreds of years in the future, if not thousands of years. Our whole society would be hundreds to thousands of years ahead. This assumes, however, that we were all able to live in some sort of peace, and we did not destroy ourselves with our equally advanced weapons of mass destruction. Therefore, seeing that these people might alter His original plan for mankind, He confused their language and scattered them all over the earth. By doing so, a lot of knowledge was lost at that time because of obvious geographical and language barriers.

See how people all over the world are related genetically? By changing 0.1 percent of their DNA information, He made them all look different. As they traveled across the oceans, guided by the stars, and established colonies all over the world, they developed different cultures and religions. Even though we look, think, and speak differently, we are all brothers.

Over time, these groups of people grew apart in every way one can think of, including physical characteristics (It reminds me of Charles Darwin's theory which, summarized, states that if different groups of the same species are separated where they do not interact with each other, but only with the environment, they then begin to look and behave differently after a certain period of time—natural selection). The Israelites turned to the golden calf, even after they physically saw the awesome power of God after the exodus from Egypt, so what makes these people different? Now we have an explanation why the pyramids of Egypt and South America, as well as other archeological sites, have so many similarities. We can also see where the technology of that era came from. There is no more mystery.

CHAPTER 14
WHO ARE WE THEN?

Now let's get back to the "pre-man" creatures. Exactly who or what were these hominids and members of the genus Homo before "modern man" came into the picture? I'll tell you! They were an advanced evolutionary species, which can be traced back hundreds of thousands of years to millions of years ago. This is one by-product of evolution that the earth was ordered to "bring forth." Some may have had quite an elaborate evolution to the point of using tools. Modern day chimpanzees use sticks to collect ants, and bones from their victims are used as weapons to attack each other. Some birds drop bones on rocks repeatedly to crack them open to get at and eat the marrow. A lot of tools that are associated with H habilis, H. erectus, and the Neandertals could have been tools designed and used by the descendants of Adam and Eve both before and after the flood. I do believe that all of these drawings found in caves all over the world were drawn by these descendants.

What a revelation! In summary, I have just shown you how the Bible correlates well with the theory of evolution of plants and animals, including the Hominids. I have also revealed to you the origin of mankind. Plants and animals underwent the process of crevolution, evolution that was created, designed, and controlled by God. Man was created or hand-made by God, in His own image and likeness. We are humans, God's offspring, as it were. That is who we really are, the human race. For me, this revelation has answered a lot of questions I had while growing up. THERE WAS

NO <u>PREHISTORIC MAN</u>! <u>THERE WAS NO MISSING LINK!</u>

As a physician and a scientist, I was torn over what I learned in school on how to examine evidence, and how to create experiments in order to come to a rational conclusion about various issues, using the scientific method. On the other hand, I have always thought of myself as very religious. For a lot of people, these two universes don't mix; it's like trying to mix oil and water. On one hand, I was faced with scientific evidence that appeared pretty convincing as to the origin of the universe and the evolution of all life here on earth. Yet, on the other hand, biology has taught me the amazing complexities of plants and animals while medicine has taught me the even more complex nature of the human body. Every time I would look around me, I would see the signature of an amazing divine creator, just as I would see the "signature" of the changing technology in the advances of the modern-day telephone. As a scientist, I had to be objective in my thinking, and thus consider the possibility of evolution.

The numerous interpretations of the Bible by the many so-called religious experts didn't make it easier for me. The Bible, even though easy to read, lends itself to many different interpretations, especially if you approach it with already pre-conceived ideas of what things mean based on what you learned as a child. When reading the Bible, read with an open mind. Everything you learned in your childhood is probably wrong. Most of all, pray before reading the Bible and ask God to give you His interpretation and truth. His will is more important than anything else. When you look at the time period it took to get us here to the twenty-first century, one can begin to imagine how omnipotent and everlasting God is, and how insignificant we are in the total picture, which is very easy to conclude. However, don't let that discourage you because all of this was made by Jesus Christ FOR US.

We are of paramount importance to God and His plan. Your relationship with God and His Son, Jesus Christ, is the most important thing that you can have in your brief life here on earth, because it is that relationship that can earn you the reward of everlasting life. The facts and evidence are before you, so do with it what you like.

CHAPTER 15
<u>CONCLUSION</u>

The conclusion to all this is very simple. There are a lot of details that are still missing to fill in the gaps in crevolution and creation. The missing information in the latter was left out intentionally. However, scientists are still working very hard to read the story of our past that the earth has yet to reveal with respect to the former. It's not an easy task, and there are many qualified and dedicated scientists out there who are painstakingly getting the job done. They and their fellow workers are literally on their hands and knees, braving the elements the earth has to throw at them. They are to be saluted. I used to think that they were crazy and misguided, but now I have a lot more respect for them, for they were on to something that we, as Christians, completely ignored and condemned. I believe that future discoveries will prove the Bible correct.

It took guts to even suggest that anything other than creation could have occurred at a time when such thoughts were considered heresy and an abomination. However, they had only part of the equation, and they were only half-right.

There are no contradictions and inconsistencies in the fact of creation as outlined in the book of Genesis. The only confusion is caused by the lack of information in the minds of the interpreters. I hope that this book has made that totally clear.

<u>Evolution did occur, and it is a fact</u>! This only applies to living things other than man. Here, the earth was allowed to produce all sorts of animals and plants. Prehistoric man, those ape-like creatures, was no more than a highly crevolved

animal species. They were all wondrous by-products that the earth was allowed to bring forth. My feeling is that man is over-interpreting bits and pieces of evidence discovered so far to prove his point of human evolution. It's probable that many of the attributes linked to these creatures are probably inaccurate and overstated to make them appear human. These prehistoric man-like beings became extinct, like many other creatures, because they were not suitable for God's purpose and plan. This evolution was controlled and guided by God, or what I call "crevolution."

Creation did occur and is a fact! Here, God did the creating Himself. He made man in His own image and likeness. Everything that was created, Adam, Eve and the Garden of Eden, as well as those specific animals that lived in the garden, was done so directly or indirectly through angels, and by Him, in His very presence. This creation did not take thousands or millions of years, but was fairly immediate. God, as you can see, has had a hand in everything that has ever lived on this earth from the very beginning.

Now you know your roots! The search is finally over. We are not animals, as some would have you believe, but human beings created in the image and likeness of the one and only true God. Unlike the rest of the numerous amounts of living beings, plants and animals, we are unique, God-like beings, shaped and created directly by God Himself (We may have some of the basic characteristics of our earthly companions, meaning animals, but that shouldn't confuse us to believe that we ourselves are animals, or evolved from them). When I say "we," I mean every human individual that has ever existed, every human being that is alive today, and every human being that is yet to come into existence. God does not look at our nationality or the color of our skin, but looks at our hearts, hearts that are, in part, His, but have been misled by Satan throughout the millennia.

I think you should read this book again! There is a lot of information in here to assimilate. I am not talking about the amount of information, but information with a twist. If you were not able to open your mind and think differently the first time around, then read the book again. You might be afraid to read this book again because it totally conflicts with what you have been taught as a child. It might totally conflict with your present occupation and beliefs. This book will probably shatter everything that you relied on to explain your very existence. Some of you may not want to accept this book as it is because your minds are not ready or willing to do so. Others would be so caught up in their own theories that to deny what they believe would discredit their fame and honor. Many are too proud and selfish to even acknowledge that they might be wrong. They would not, or cannot, turn back because to do so would mean a life wasted on nothing. I tried to get permission from a museum to use their images from their collection of "prehistoric man" in this book, but I was turned down for what I believe was interpreted as creation propaganda. What's the big deal?

Our lives are never a waste, even if we learn the truth at the very end. Whether we accept God and Christ at the very beginning, or at the very end, our reward will be equal. I'm talking about true acceptance. He has promised that to us. We shall all be given the opportunity to know God, whether it is in this life or in the world to come. Everyone, the dead and the living, will be given a chance to choose the direction he or she will take. He will make sure that every opportunity will be given to mankind before any judgment is made. After all, God sent His only begotten Son to die for us. He does not want any of us, great or small, to perish.

Then how important are we to Him? Would you put your child at risk to save this modern human race? I'm talking about all of humanity, not just a selected few. I didn't think so! He sent His only begotten Son to die a miserable death just for us, the whole human race. That is how much He loves us. His love for us will never cease. That is what I call an equal opportunity God.

This century heralds a new era of the unknown. The optimists predict a life of convenience, luxury, innovations, and amazing technologies beyond our imagination. The pessimists, on the other hand, predict doom and gloom, as our society deteriorates possibly to extinction. The century hasn't started out as some have planned, and if things do not change, the pessimists may take the lead. Many of us today are frightened and concerned about our own security, as well as that of our children. For those who, at one time or the other, had children, think how scared you were when one of them got really sick? Remember the panic? That is what many of us feel today when at first we thought we were in control of our lives, but soon found out that we were in control of nothing. A lot of mental health care professionals propose numerous remedies to help overcome this feeling, but I'm afraid that it is not lasting. They are also scared themselves, but have to appear otherwise. Things much worse than what occurred on 9/11/01 are going to happen in the near future! Where does your trust and faith lie?

Is God an important part of your life? If not, He should be. Give Him the chance to be in your life. He is just and fair, but you just do not know it. Many of you can't see that right now. For those of us who pass the test, we shall become just like God and be placed higher than the angels, and be taken care of for all eternity by God and the angels. And that is one of His promises. He is the <u>one</u> and <u>only</u> true, still-living God.

USEFUL DEFINITIONS

1. Chromosomes—Found in the nuclei of cells and made up of long strands of DNA.

2. Mt-DNA—DNA material found in cytoplasmic bodies called mitochondria

3. Deoxyribonucleic Acid (DNA)—The helical molecule located in the nuclei of cells and which carries the genetic code

4. Y-chromosome—This type of chromosome imparts the male gender to the human species; Females have two -XX chromosomes, and males have XY chromosomes.

5. Giants—These are the offspring of the supernatural beings, angels, and the daughters of man at the time of Noah.

6. Malaria—A disease spread by the bite of a mosquito, Anopheles, and caused by a protozoa, Plasmodium

7. Sickle Cell Anemia—A genetically inherited illness that causes the red blood cells to form in the shape of a sickle. As a result, the sickle cells have a shorter lifespan and break down more rapidly, thus resulting in anemia.

8. Tsunami—Large tidal waves.

ABOUT THE AUTHOR

The author is presently a general practice physician in the state of California, and has practiced medicine for 18 years. He is a graduate of the Oregon Health Sciences University in Portland, Oregon. Never did he think that he would author a book of such immense spiritual consequence. A Christian by faith, he was misled in his youth by numerous Christian religions down a path of blind faith. His training as a physician only strengthened his belief in God, but underneath lay turbulent unanswered questions with respect to evolution and its missing link with creation. His faith in God finally opened his eyes in what he refers to as an unprecedented book.

ABOUT THE BOOK

This book you have in front of you is like nothing that you have ever experienced, or ever will, when it comes to the topic of evolution vs. creation. The learned theologians and the intellectual archeologists and evolutionists couldn't figure it out! This book is finally going to put the issue of our origin to rest.

This unique book combines the pages of Genesis with the pages of archeological history over the past 150 years. It tries to unite the creationists and the evolutionists in an effort to finally show that both evolution and creation actually occurred. Finally revealed is convincing evidence that both evolution and creation are FACTS. Nowhere else can anyone find such evidence, and believe it or not, it's been sitting under our very noses for thousands of years.

Evolutionary theories based on archeological evidence had the story half right. The other half is in the first two chapters of Genesis, hidden for all these years. It took the discoveries of the scientists to highlight the secrets in Genesis. Based on both bodies of evidence, one can conclude that man was created, while everything else underwent the process of evolution over millions of years.

PREVIEW

The part of the book that I would preview would be the chapter labeled conclusion.

REFERENCES

1. http://www.ecotao.com/. Human Evolution by Dr. Laurence Evans.
2. http://www.versiontech.com/.Origins Of Humankind News, New Out-Of-Africa Theory   Unveiled by Larry O'Hanlon.
3. http://www.pathlights.com/.Ancient Man-2
4. Of Mud and the Divine: Creation Myths of the Middle East: Part Two by Ewa Wasilewska.
5. Charles Darwin and the Theory of Evolution; http://www.edlibrary.com
6. Darwin on The Origin of Species: by George Landow.
7. Creation of a Cosmology: Big Bang Theory: http://ssscott.tripod.com/ Big Bang.html
8. The Geological Time Machine, http:/www.ucmp.berkeley.edu
9. Pictures Of Charles Darwin, http:/academy.d20.co.edu
10. www.biozone.co.uk
11. www.wsu.edu
12. The Holy Bible: The Good News Bible, the Bible in today's English version.

www.ingramcontent.com/pod-product-compliance
Lightning Source LLC
Chambersburg PA
CBHW022026170526
45157CB00003B/1377

* 9 7 8 1 4 1 8 4 4 9 7 0 4 *